Food Plant Safety
UV Applications for Food and Nonfood Surfaces

Food Plant Safety
UV Applications for Food and Nonfood Surfaces

Tatiana Koutchma
Agriculture and AgriFood Canada,
Guelph Food Research Center,
Guelph, Ontario, Canada

AMSTERDAM • BOSTON • HEIDELBERG • LONDON
NEW YORK • OXFORD • PARIS • SAN DIEGO
SAN FRANCISCO • SINGAPORE • SYDNEY • TOKYO

Academic Press is an imprint of Elsevier

Academic Press is an imprint of Elsevier
32 Jamestown Road, London NW1 7BY, UK
The Boulevard, Langford Lane, Kidlington, Oxford, OX5 1GB, UK
Radarweg 29, PO Box 211, 1000 AE Amsterdam, The Netherlands
225 Wyman Street, Waltham, MA 02451, USA
525 B Street, Suite 1900, San Diego, CA 92101-4495, USA

Copyright © 2014 Elsevier Inc. All rights reserved.

No part of this publication may be reproduced or transmitted in any form or by any means, electronic or mechanical, including photocopying, recording, or any information storage and retrieval system, without permission in writing from the publisher. Details on how to seek permission, further information about the Publisher's permissions policies and our arrangement with organizations such as the Copyright Clearance Center and the Copyright Licensing Agency, can be found at our website: www.elsevier.com/permissions

This book and the individual contributions contained in it are protected under copyright by the Publisher (other than as may be noted herein).

Notices
Knowledge and best practice in this field are constantly changing. As new research and experience broaden our understanding, changes in research methods, professional practices, or medical treatment may become necessary.

Practitioners and researchers must always rely on their own experience and knowledge in evaluating and using any information, methods, compounds, or experiments described herein. In using such information or methods they should be mindful of their own safety and the safety of others, including parties for whom they have a professional responsibility.

To the fullest extent of the law, neither the Publisher nor the authors, contributors, or editors, assume any liability for any injury and/or damage to persons or property as a matter of products liability, negligence or otherwise, or from any use or operation of any methods, products, instructions, or ideas contained in the material herein.

British Library Cataloguing-in-Publication Data
A catalogue record for this book is available from the British Library

Library of Congress Cataloging-in-Publication Data
A catalog record for this book is available from the Library of Congress

ISBN: 978-0-12-416620-2

For information on all Academic Press publications visit our website at **store.elsevier.com**

This book has been manufactured using Print On Demand technology. Each copy is produced to order and is limited to black ink. The online version of this book will show color figures where appropriate.

www.elsevier.com • www.bookaid.org

CONTENTS

Chapter 1 Introduction 1

Chapter 2 Basic Principles of UV Light Generation 3
2.1 UV Rays: Radiation or Light 3
2.2 UV Light Sources 4
2.3 Commercial Manufacturers of UV Systems for Sanitation
     of Food Facilities 11

Chapter 3 UV Disinfection of Air, Water, and Surfaces 15
3.1 Air Treatment 15
3.2 Water Treatment 19
3.3 Disinfection of Nonfood Contact Surfaces 22
3.4 Disinfection of Food Contact Surfaces 23

Chapter 4 Case Studies of UV Treatment of Food Surfaces 31
4.1 Fresh Produce 31
4.2 Fresh-Cut Produce 33
4.3 UV- and PL-Based Processing Aids for Meats 35
4.4 Shell Eggs 36
4.5 Seafood 37
4.6 UV Equipment for Treatment of Food Surfaces 38

Chapter 5 Principles of UV Surface Process Development 41
5.1 Factors Affecting Interaction Between UV Light
     and Surface of Materials 41
5.2 Determination of UV Dose 41
5.3 Environmental Assessment 43
5.4 Regulatory Status 45

Chapter 6 Conclusions 47

References 49

CHAPTER 1

Introduction

During manufacturing process, ingredients, raw materials, semifinished and finished products can be exposed to microbiological cross-contamination from the air, water, and surfaces in food and storage facilities. The traditional approach to controlling such contamination has been to target specific sites within the manufacturing environment with cleaning and disinfection regimes. Ultraviolet (UV) light is an economical intervention toward improved hygiene control measures in the food industry. Sanitation, decontamination, disinfection, and oxidation with UV light is a versatile, environmental-friendly technology, which can be used in the food production and storage facilities to reduce microbial contamination and consequently to improve safety of finished products. UV light has been historically used for purification of air, disinfection of water and wastewater. In the last decades among other novel techniques, UV technology started to emerge in food processing practices because of its broad antimicrobial action, low cost, and nonthermal, purely physical nature. Despite UV light being known for its surface character, the main interest was focused at the application of UV light as an alternative means of food preservation of food fluids with a broad range of UV light transmission (UVT). There is limited research and commercial practices to use UV as an additional step or/and as a processing aid to disinfect nonfood contact and food contact surfaces to reduce risk of contamination and cross-contamination. This book is aimed to review benefits of continuous UV light and pulsed light for the applications to improve safety at the food plants and provide perspective to food processors on applications of those benefits for their production facilities. The current and perspective applications of UV light for air purification, water disinfection, and surface treatments in food processing and storage facilities will be reviewed among with available UV sources and commercial equipment.

CHAPTER 2

Basic Principles of UV Light Generation

2.1 UV RAYS: RADIATION OR LIGHT

UV radiation is defined as the portion of the electromagnetic spectrum between X-rays and visible light. The wavelength for UV light diapason ranges from 100 to 400 nm. This range may be further subdivided into: UVA (315–400 nm) normally responsible for tanning in human skin; UVB (280–315 nm) that causes skin burning and can lead to skin cancer; UVC (200–280 nm) called the germicidal range since it effectively inactivates bacteria and viruses. Vacuum UV range (100–200 nm) can be absorbed by almost all substances and thus can be transmitted only in a vacuum or through certain gases like N_2. Radiation from UV light and the adjacent visible spectral range as well as other less energetic types are termed nonionizing radiation. In contrast, ionizing radiation, which includes X-rays, gamma rays, and ionizing particles (beta rays, alpha rays, and protons), is capable of ionizing many atoms and molecules. The absorption of nonionizing radiation, however, leads to electronic excitation of atoms and molecules.

Light is emitted from the gas discharge at wavelengths dependent upon its elemental composition and the excitation, ionization, and kinetic energy of those elements. UV photons have low penetrating power, because the inherent energy of photons is low in comparison with ionizing radiation and belongs to nonionizing radiation. Absorption of nonionizing radiation leads solely to electronic excitation of atoms and molecules as opposed to ionizing radiation that can lead to ionizing effects. However, the excitation energy of UV photons is much higher than the energy of thermal motions of the molecules at physiological temperatures. The latter is of the order of Boltzman's constant times the absolute temperature that at 27°C amounts to 0.026 eV/molecule (0.60 kcal/mole) in contrast to the 3.3–6.5 eV/molecule (75–150 kcal/mole) available from UV absorption.

2.2 UV LIGHT SOURCES

The gas discharges are responsible for the light photons emitted from UV lamps.

UV light transfer phenomenon is defined by the emission characteristics of the UV source along with long-term lamp aging and absorbance/scattering of the product. The commercially available UV sources include low- and medium-pressure mercury lamps (LPM and MPM), excimer (EL) and pulsed light (PL), and light-emitting diodes (LEDs). LEDs and ELs are monochromatic sources whereas emission of MPM and PL is polychromatic. Consequently, performance of UV system depends on the correct matching of the UV source parameters, its wavelength and output power, to the demands of the UV application and its characteristics.

2.2.1 Mercury Lamps

The mercury vapor UV lamp sources have been successfully used in air and water treatment for nearly 50 years and well understood as reliable sources for other disinfection treatments that benefit from their performance such as long lifetime, low cost, and quality. Typically, three general types of mercury UV lamps are used: low-pressure (LPM), low-pressure (high-output) amalgam (LPA), and medium-pressure (MPM). These terms are based on the vapor pressure of mercury when the lamps are operating.

LPM lamps are operated at nominal total gas pressures of $10^2 - 10^3$ Pa that corresponds to the vapor pressure of mercury at temperature of 40°C with no additional cooling system required. The emission spectrum of LPM is concentrated at the resonance lines at 253.7 nm (85% of total intensity) and 185 nm. The wavelength of 253.7 nm is considered as most efficient in terms of germicidal effect since photons are absorbed most by the DNA of microorganisms at this specific wavelength. Light with a wavelength below 230 nm is most effective for the dissociation of chemical compounds. The photons with the wavelength of 185 nm are responsible for ozone production and the combination of both wavelengths is a very effective means for photochemical air treatment. LPM sources provide a high efficiency in inactivating microbial cells and a number of other advantages such as lifetime up to 9000–12,000 h. Standard LPM quartz lamps are available as "ozone-free" and ozone-generating versions, depending on different transmittance properties of quartz glass. For "ozone-free" lamps doped fused quartz is used (TiO_2 to cut transmittance below 235 nm). The US Food and Drug Administration (FDA) regulations approved the use of LPM lamps for water and surface treatment, and they have already been successfully commercialized in food industry (US FDA, 2000a).

More powerful options exist for the future once validated, such as high-intensity lamps and MPM lamps. LPA lamps that contain a mercury amalgam were developed and incorporated into disinfection applications. The lifetime of amalgam lamps can be up to 12,000 h. LPA lamps offer high UVC output, excellent lifetime, good UV efficiency, high temperature stability and enable operation in high ambient temperatures. This combination makes them the right choice for compact, efficient, and economic disinfection and advanced oxidation systems.

MPM lamps are operated at a total gas pressure of 10^4–10^6 Pa. Compared to the LPM lamps, the coolest possible temperature of the MPM is about 400°C, whereas it goes up to 600°C and even 800°C in a stable operation and active cooling can be required. The emission spectrum of MPM covers wavelengths from about 250 nm to almost 600 nm, which results from a series of emissions in the UV and in the visible ranges. MPM lamps are not considered to be useful for targeted germicidal treatment. By varying the gas filling, doping, and the quartz material, the spectrum as well as the radiation flux of the UV lamps can be varied and matched to suit specific food processing applications, especially for oxidation or photo degradation.

Ballasts power all lamps. They provide the starting electrical voltage to ionize the gas in the UV lamp and then limit the current to the nominal level. Lamp ballasts can be either magnetic or electronic. The main components of mercury UV sources include power supply, ballasts for regulated power supply, support structure for the lamps and quartz sleeves that are shown schematically in Figure 2.1.

Germicidal UVC lamps must be fabricated from specialized quartz rather than ordinary glass. The lamp is installed inside of quartz sleeve with Teflon coating less than 1 mm thickness. Teflon coating allows 90–95% of the UVC germicidal energy to go through, however it will contain the pieces in case of lamp breakage and be resistant to lamp

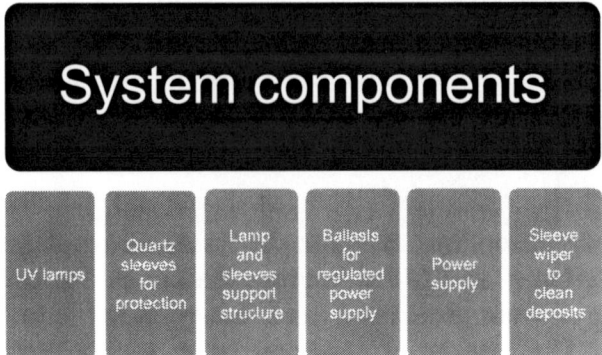

Figure 2.1 Typical components of UV light mercury sources.

operating temperatures. UV lamps require replacement every year, and the quartz sleeve that houses the lamps every 2–3 years. Wiper seals typically require annual replacement. The routine downtime per system is typically 1–2 h, and most manufacturers provide factory training for customers' technical staff if required.

2.2.2 PL Lamps

PL has a broader range of the wavelengths of 170–1000 nm similar in wavelength composition to the solar light and often combines UV photons, visible and infrared (IR) light at energies of 0.01–50 J/cm^2. The UVC part of the spectrum is the most important for microbial inactivation.

PL is generated using a capacitor and delivered in several flashes of light per second, allowing fast throughput of product, and low energy usage. The intensity of PL is about 20,000 times the intensity of UV light. The efficacy of pulsed flash lamps is potentially greater than continuous sources due to high intensity, broader spectrum, instant start, and robust packaging with no mercury in the lamp. In this technology, alternating current is stored in a capacitor and energy is discharged through a high-speed switch to form a pulse of intense emission of light within about 100 ms. Xenon lamps are commercial sources of PL that require air or water cooling for the operation. The short pulse width and high doses of the PL source may provide some practical advantages over continuous UV (cUV) sources in those situations where rapid disinfection is required. Other advantages of PL treatment are the lack of residual compounds, and the absence of applying chemicals that can cause ecological problems and/or is potentially harmful to humans. Sample heating is perhaps the most important limiting factor of PL for practical applications. Heat can originate from the absorption of IR part of PL by the food or by lamp heating. Another disadvantage of PL treatments is the possibility of shadowing occurring when microorganisms readily absorb the rays.

The following trend of susceptibility of microorganisms to PL in decreasing order was reported: Gram-negative bacteria, Gram-positive bacteria, and fungal spores. The color of the spores can play a significant role in fungal spore susceptibility. *Aspergillus niger* spores are more resistant than *Fusarium culmorum* spores, which could be because the pigment of the *A. niger* spores absorbs more in the UVC region than that of *F. culmorum* spores, protecting the spore against UV (Gomez-Lopez et al., 2007).

The factors determining the interaction of PL exposure with foods are in some extent similar to UV light treatments. The critical factor affecting PL treatments is fluence incident on the sample. The energy emitted by the flash lamp is different from the energy incident on the sample. Factors such as distance from light source to target and propagation vehicle (air, water, or other fluids, dust particles) affect the level of energy than ultimately reach the target. The inactivation efficacy of PL is higher when treated products are closer to the lamp. Food composition also affects the efficacy of the decontamination by PL (Gomez-Lopez et al., 2007). High protein- and fat-containing food products have little potential to be efficiently treated by PL. Vegetables, on the other hand, could therefore be suitable for PL treatment.

Figure 2.2 shows the normalized spectra of cUV sources such as LPM, MPM, and pulsed sources (PL). Individual spectra are not comparable on a UV intensity basis but are comparable on a spectral basis regarding which wavelengths dominate the respective wavelength outputs.

2.2.3 Eximer Lamps

The ELs are considered as the alternated UV lamps as their specific feature of narrow emission band depends on the choice of rare gas and/or halogen (e.g., KrCl*, $\lambda = 222$ nm; XeBr*, $\lambda = 282$ nm) (Sosnin et al., 2006; Koutchma, 2009). Today, there are several different excimer

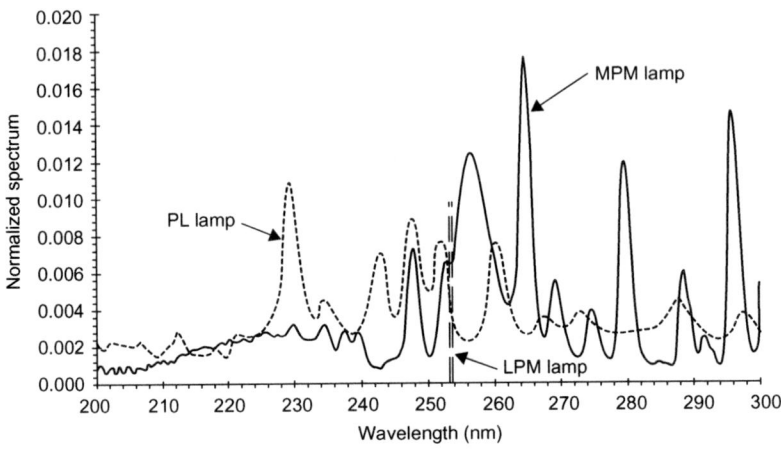

Figure 2.2 Normalized spectra of LPM, MPM, and PL sources and intensity.

combinations, which can produce UV radiation in the wavelength range between 120 and 380 nm. Typically, ELs have a coaxial geometry with an inner and outer electrode and a double cylindrical quartz body. ELs also have an advantage of extremely low heat output and are able to operate at much lower surface temperatures. Thus, they can provide an advantage by avoiding fouling behavior by liquid foods. Most significant is the higher specific UVC-flux per unit plasma length of the XeBr* lamps. Further, ELs are mercury free and instant with no warm up time. These ELs, however, suffer from a low UVC efficiency of $\sim 8\%$ compared to 35% for LPM and LPA lamps. Another drawback to date is the high investment costs for lamps and power supplies. Warriner et al. (2000) demonstrated that UV-excimer light was effectively used for sterilization of the packaging carton surfaces.

2.2.4 Light-Emitting Diodes

In recent years, UV LEDs have been developed with the following many advantages: low cost, energy efficient, long life, easy control of emission, and no production of mercury waste. The wavelength of the commercial UV LED that is in the germicidal range of 260–400 nm can enable new applications in existing markets as well as open new areas. An LED is a semiconductor device that emits light when carriers of different polarities (electron and holes) combine generating a photon. The wavelength of the photon depends on the energy difference the carriers overcome in order to combine. The example of UV LED system that operates between 210 and 365 nm is the one formed by aluminum nitride (AlN), gallium nitride (GaN), and intermediate alloys. Unlike traditional light sources, whose output wavelength is fixed, UV LEDs can be manufactured to operate at the optimum wavelength for the application: 265 nm is widely recognized as the peak absorption of DNA; however, it was demonstrated that the peak disinfection efficacy of Escherichia *coli* in water occurs at 275 nm.

UV LEDs also switch on and off instantly and can actually be pulsed without any detriment to lifetime, making them more user-friendly and safer for the operator. The design rules for UV LEDs open new opportunities of what can be disinfected: we are no longer limited to a long tube, but can mount the LEDs on flat panels, flexible circuit boards, and the outside of cylinders; the options are almost endless. Currently, UV LEDs are commercially available at research grade in limited quantities and their lifetime reach on the order of 200 h. It is very likely that in the near future, UV LEDs will carry out many applications that today make use of mercury lamps. Crystal IS, Inc., Green Island, NY (CIS, NY, http://www.cisuvc.com/) has started providing high-quality AlN native substrates with a dislocation density in the order of $1000/cm^2$. In the past 5 years, it has also started the development of UV LED devices. A single and package of LEDs and the emission spectra of the individual LEDs are shown in Figure 2.3.

2.2.5 Efficiency of UV Light Sources

The UV sources generally convert their input watts into usable UVC watts. There is a number of most important technical characteristics associated with UV lamps that can be used to compare operating performance of the cUV, PL, and LED sources. This includes operating temperature, electrical input (W/cm), germicidal UVC output or UVC flux (W/cm), electrical to germicidal UV conversion efficiency (%), arc length or the length of the lamp being used, UV irradiance or intensity (mWs/cm^2) at a specific distance, and rated lifetime (h). The specific electrical loading in the glow zone, expressed in watts per centimeter, is also used to compare efficiency typically between 0.4 and 0.6 W (e)/cm. The linear total UV output of the discharge length for lamps appropriate for use in disinfection is in the range of 0.2–0.3 W (UV)/cm. This means that the UV efficiency generally designed by total W (UV) output versus W (e) input is between 0.25 and 0.45. The energy losses are mainly in the form of heat.

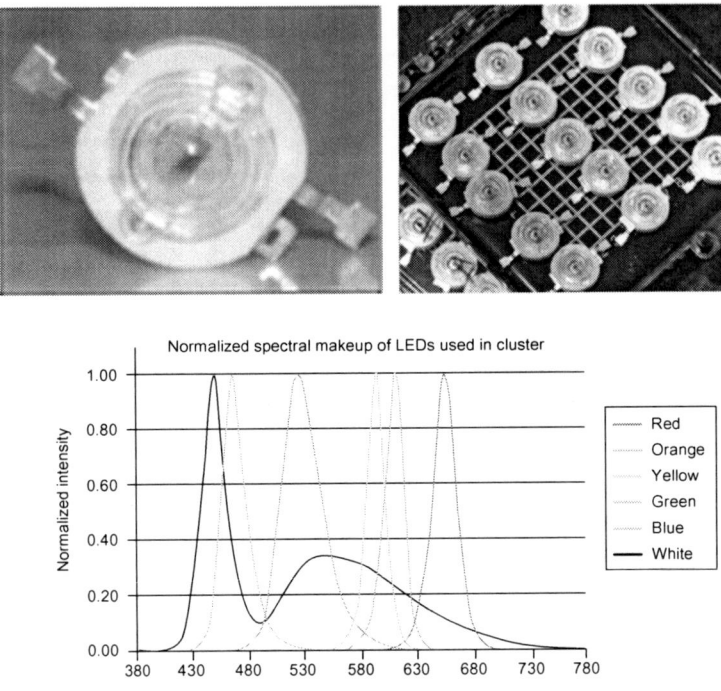

Figure 2.3 Single LED and packaged LEDs and their spectra.

For LPM lamps, the UVC proportion of the UV light wavelengths emitted is in the range of 80–90% of the total UV power emitted. Approximately 40% of the electrical power is converted into UVC radiation at 254 nm. The material of the envelope and aging of LPM lamps also influence the emitted intensity and efficiency. The preferred envelope material for LP is fused quartz. Due to the generally low wall temperature, it is also possible to use softglass (sodium–barium–glass). The UV-softglass does not transmit at 185 nm, hence all these lamps are "ozone-free" lamps. In contrast to softglass lamps, standard quartz lamps are available in both "ozone-free" (G, GPH lamps) and ozone-generating (G...VH, GPH...VH—VH stands for very high ozone) versions. Typically, both the specific UVC-flux per unit arc length and the UVC efficiency are higher for the fused quartz types. An initial drop in emission yield occurs during the first 100–200 h of operation. However, after this period of time, the emission is stable for months. Aging is caused by two factors: solarization of the lamp wall material on the one hand and blackening due to deposits of sputtered oxides from the electrodes.

Table 2.1 Efficiency of Continuous LPM, MPM, and EL Lamps; PL Lamps; and LEDs

UV Source	Electrical Efficiency, %	UV Efficiency, %	UV Intensity, W/cm^2	Lamp Surface T, °C	Lifetime, Hours	Output Spectrum
LPM	50	38	0.001–0.01	40	12,000	Monochromatic, 253.7 nm
MPM	15–30	12	12	400–1000	5000	Polychromatic, 200–400 nm
PL Xenon	15–20	9	600	1000–10,000	800	Polychromatic, 100–1000 nm
EL	10–35	10	150	Ambient	1000	Monochromatic, 200–400 nm
LED	1–4	NA		50–60	10,000	Monochromatic, 200–400 nm selective

A comparison of the various features and efficiency characteristics of cUV, PL, and LEDs sources is given in Table 2.1. It is evident that no single lamp technology will represent the best source for all in-plant or food applications. However, situation-specific requirements may dictate a clear advantage for a given process technology. Lamp prices are generally higher for lamps with higher UV fluxes. Mainly, the characteristics and cost determine the optimum UV source for a specific application. Special technologies lamps such as PL, EL, and LEDs are promising due to different spectral bands or specific wavelength that they can provide for different applications. Each application requires its specific UV treatment parameters such as, radiation spectrum, UV output, light tube temperature, illumination length, and lamp geometry are precisely adjusted to the respective operating conditions. More research is needed to establish their suitability for food processing applications.

2.3 COMMERCIAL MANUFACTURERS OF UV SYSTEMS FOR SANITATION OF FOOD FACILITIES

There is a large number of manufacturers of UV equipment suitable for food processing facilities and rooms using various UV sources including LPM, LPA, MPM, EL, and PL that are typically supplied with automated, on line wipers that keep lamps free from fouling.

Heraeus Noblelight (Hanau, Germany, http://www.heraeus-noblelight.com/en/home/noblelight_home.aspx) is a major player on the market of manufacturing UV sources and UV disinfection equipment that supplies UV lamps and process solutions for numerous fields of application which are precisely tuned to the plant and the process. This saves energy, maintenance, and operating costs and enhances quality. Heraeus offers classic LPM sources of various lengths, shapes, basis, and connections, EL for surface treatment, arc and flash PL lamps, and LEDs that can be optimized for individual air, water, and surfaces treatment applications.

sterilAir (Weinfelden, Switzerland, http://www.sterilair.com/en/) has been offering developing and manufacturing UVC disinfection devices and components for solution-orientated hygiene concepts since 1939. sterilAir UVC technology is primarily used for the disinfection of air, surfaces, and fluids in the food industry, air-conditioning equipment, building construction technology, laboratories, medicine, and livestock breeding.

Sanuvox Technologies, Inc. (Montreal, QB, http://www.sanuvox.com/en/index.php#about) is also focused on designing a line of residential and commercial UV air purification systems that would address indoor air quality (IAQ) issues.

American Ultraviolet (Lebanon, IN http://www.americanultraviolet.com/germicidal-ultraviolet.cfml) manufactures UVC germicidal fixtures for commercial, healthcare, and residential applications to reduce harmful food pathogens, increase efficiency, prolong equipment life, and improve IAQ.

Radiant Industrial Solutions, Inc. (Houston, TX, www.radiantuv.com) has been provider of UV-based disinfection solutions since 1989. Radiant UV offers a complete line of products for air, water, and surface disinfection, as well as targeted application-specific solutions in food and beverage, pharmaceutical, semiconductor, medical, and curing markets.

Healthy Environment Innovations (HEI, Dover, NH, http://www.he-innovations.com/index.html) specially designs novel monochromatic EL sources with specific wavelength at 222 and 282 nm in germicidal range that are called Far UV and Far UV + for emitting single wavelength light based on the excimer discharge. These EL UV sources have a potential to significantly enhance the microbial inactivation on the contact surfaces and foods, in the rooms and in the air and destroy toxic chemicals such as mycotoxins by emitting photons with higher energy or matching emission spectrum to the peak of material absorption. Sterilray™ Technology is capable of destroying many of the pathogenic bacteria and viruses much faster and more effectively than chemical disinfection methods.

The pioneer company producing PL equipment for disinfection was PurePulse Technologies, Inc. (San Diego, CA), a subsidiary of Xenon Corp. Applications included water purification systems and virus inactivation systems for biopharmaceutical manufacturers. Nowadays, there are three commercial companies producing pilot-scale and commercial systems based on PL: Claranor from France (http://claranor.com), SteriBeam Systems from Germany (http://www.steribeam.com/), and Xenon Corporation from the United States (http://www.xenoncorp.com/). Claranor develops complete sterilization solutions for food and beverage industrial applications, including caps and cups sterilization prior to filling/bottling. The detailed information regarding the devices for industrial applications can be found at the websites of these companies.

CHAPTER 3

UV Disinfection of Air, Water, and Surfaces

3.1 AIR TREATMENT

Clean, fresh air is the basis in the industrial food production. Microorganisms in the air, such as viruses, bacteria, yeasts, and fungi, can contaminate raw materials and intermediate products and spoil finished products during their processing and packaging. LPM sources are used very successfully in these applications, for disinfection in air intake ducting and storage rooms and to ensure air of very low microbial content in the production areas. Short-wave UV radiation at 185 nm produces ozone from the oxygen in the ambient air that is activated for the oxidation process. UV oxidation breaks down pollutants in the exhaust air. For providing clean air in the sensitive manufacturing food facilities, a combination of filters and UV light has been recommended. Basically, two applications of UV light are becoming common. In one, the moving air stream is disinfected in much the same manner as with a water system. In the other application, stationary components of the system such as air conditioning coils drain pans and filter surfaces are exposed to UV helping to prevent mold and bacteria growth or to disinfect the filter to aid in handling. The UV transmittance (UVT) in air is higher than in water and, therefore, the number of lamps required in a large duct is quite reasonable. Common airborne virus and bacteria are readily deactivated with UV. Fungi (molds and spores) require much higher doses due to their higher UV resistance that depends on air temperature and humidity. In the moving air stream, high-wattage lamps are used, usually without a quartz sleeve. UV lamp fixtures can be placed in such a manner as to completely irradiate surfaces where bacteria and mold might collect and grow. Mathematical modeling software and bioassay testing have been developed, to allow efficient design and validation of these systems. Low operating costs and reasonable equipment costs can make UV very cost effective.

sterilAir offers air recirculation units to deal with air volumes of 180–500 m³/h, wall-mounted systems and ceiling-mounted emitters,

Food Plant Safety. DOI: http://dx.doi.org/10.1016/B978-0-12-416620-2.00003-5
© 2014 Elsevier Inc. All rights reserved.

Figure 3.1 sterilAir recirculation system. (Credit to sterilAir, Weinfelden, Switzerland.)

HVAC-modules that can be retrofitted and incorporated in already existing air conditioning equipment and components. As examples of air purification solutions, sterilAir air recirculation systems (Figure 3.1) disinfect room spaces with either a low convection or a direct emission should be prevented.

The example of sterilAir industrial UV light based duct air purification system is shown in Figures 3.2 and 3.3.

Radiant UV (Houston, TX) focused on developing solutions to use UV as a control measure for air quality in the areas with targeted microbial levels. The commercial solutions can be used in air circulation and recirculation systems, product packaging areas, product transport, wet air process control, and air to positive pressure directional flow. In 2010, UV disinfection was introduced as a control method for plant air with microbial targets such as compressed air. Key process attributes

UV Disinfection of Air, Water, and Surfaces   17

Figure 3.2 streilAir duct system. (Credit to sterilAir, Weinfelden, Switzerland.)

include compressed air volume at the pressure of 120 pound per inch (PSI), effective against *Listeria* target pathogen. Radiant UV dose can be measured and displayed on the remote monitor (Figure 3.4).

American Air and Water Company also offers commercial air purifiers that can be installed in a duct system, on a ceiling, walls, and floor. Technical specifications and description can be found at http://www.americanairandwater.com. Mobile UV systems also can be used for air treatment in the processing rooms after cleaning. Fuller UV, Inc. (http://www.fulleruv.com/ultraviolet_and_ozone_lamps.html) offers simple wall fixtures made of stainless steel with UVC lamps in protective jackets.

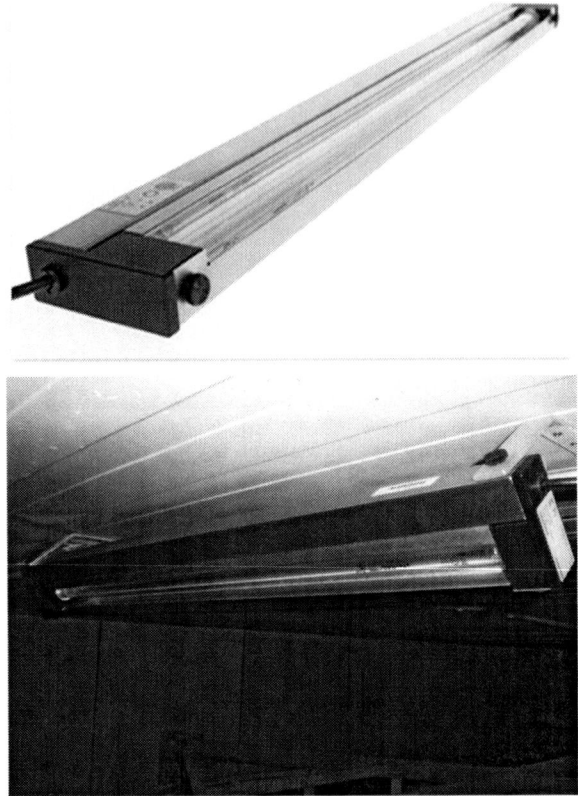

Figure 3.3 streilAir wall-mounted air cleaning UV system. (Credit to sterilAir, Weinfelden, Switzerland.)

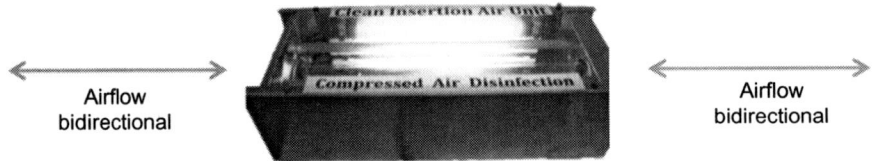

Figure 3.4 Radiant UV compressed air UV disinfection system. (Credit to Radiant UV, Houston, TX.)

In air conditioning systems, UV lamps are used in washer tanks (encapsulated lamps), cooling coils or to disinfect the air stream directly. UV air oxidation is used for odor removal (in sewage plants, rest rooms, hotels, restaurants, catering, caravan trailers, and cars), grease destruction in kitchen hoods, and industrial exhausts. For air temperatures below 40°C, standard ozone-generating LPMs are utilized. For higher temperatures, it is essential to use ozone-generating LPA lamps.

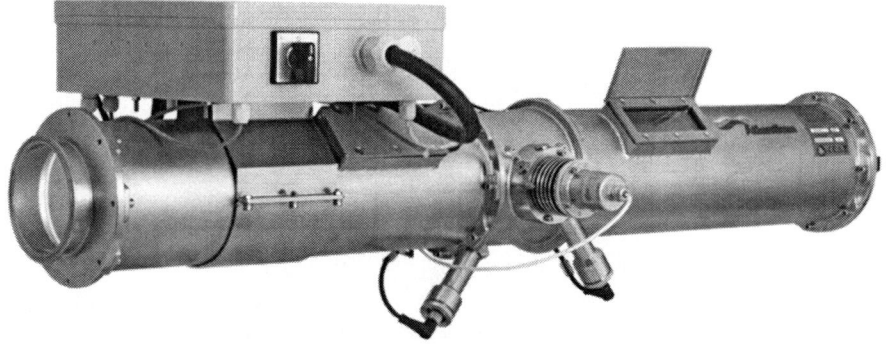

Figure 3.5 Atlantium UV system with optical sensors to monitor UV dose delivery. (Credit to Atlantium Technologies, Israel.)

3.2 WATER TREATMENT

Control of microorganisms in industrial process waters is often necessary to maintain quality of the product or process. The food industry is a large-volume consumer of water, and the potential for reuse or recycling of fruit processing water represents an attractive economic and sustainable benefit to the industry. A combination of UV light and ozone is a powerful oxidizing action to reduce microbial load and the organic content of water to very low levels. Standard quartz LPMs and LPAs are used across the whole range of water applications for disinfection and/or oxidation processes (e.g., drinking water, wastewater, domestic water, ground water, industrial water, and ultra-pure water) with small and medium high flow rates. High UV-flux lamps, such as amalgam and MPMs, are widely used also with the exception of small residential drinking water systems. Water treatment systems with LPAs are highly energy and cost efficient. MPMs are used if space efficiency is of primary consideration (Schalk et al., 2006).

Atlantium Illuminating Water Technologies (Har Tuv, Israel, http://www.atlantium.com/en/home.html) offers UV system for water treatment in beverage, beer, and dairy processing using MPM lamps. The UV unit is equipped with two integrated sensors that can continuously monitor UV lamps intensity and water quality for UV transmittance and thus enabling automatic dose adjustment according to conditions in real time. The Atlantium UV system shown in Figure 3.5 has been validated to meet EPA requirement for 4 log reduction of Adenovirus. The 2009 Pasteurized Milk Ordinance (PMO) included criteria for the use of UV to provide disinfected and pasteurized equivalent water in dairy processing. Atlantium's

Figure 3.6 Commercial SurePure Unit (SP 40). (Credit to SurePure, Inc., Zug, Switzerland.)

systems are validated to the criteria required by the PMO such as a 5 log bacterial reduction and 4 log virus reduction, programmed, and installation-ready. The system replaced water heat pasteurization to rinse curds in the cottage cheese line and can be used for product water, ingredient water, and Cleaning-IN-Place (CIP) disinfection.

SurePure, Inc. (Zug, Switzerland, http://www.surepureinc.com) offers new UV photopurification technology. Fluid or processing water is pumped through the so-called "turbulators" that are made as a corrugated spiral tube located between an inlet and outlet chambers. The LPM source is mounted in a quartz sleeve inside the chamber (Figure 3.6). This unique design of the turbulators combined with the bactericidal action of UVC light ensures high efficacy against food pathogenic and spoilage microflora in fluids characterized by low UV transmission and wide range of viscosities. The delivery of UV dose to achieve specific microbial reduction can be accomplished by changing a number of single turbulators. Commercial SurePure systems with 6, 10, 20, or 40 turbulators can treat a variety of food ingredients, raw and finished products without destroying essential nutrients, but often enhancing their functional properties. Every installation requires its own bespoke solution—indicative

Figure 3.7 AQD-PVC series—UV systems with plastic vessels from sterilAir. (Credit to sterilAir, Weinfelden, Switzerland.)

commercial applications. The use of SurePure turbulent UV process in brewing industry that is surprisingly water intensive industry can result in cost savings due to water reduction for applications for D-water treatment, wastewaters, and steep water in malting plants. Additionally, the turbulent flow of the fluid over the lamps ensures a foul-free, self-cleaning system and provides more savings in water consumption. SurePure UV purification was successfully applied to reduce *Alicyclobacillus acidoterrestris* spores inoculated into tap water, used wash water from a fruit concentrate manufacturing facility. In water inoculated with *A. acidoterrestris* spores, a 5.3 $\log_{10}$ reduction of the alicyclobacilli was achieved after a UV dosage of only 305 J/L, resulting in no viable spores. The UV treatment method was shown to be capable of reliably achieving in excess of a 4 $\log_{10}$ reduction (99.99%) after 500 J/L of applied UVC dosage in *A. acidoterrestris* inoculated in used fruit juice concentrate factory wash water (Groenewald et al., 2013).

sterilAir flow-through photoreactors are designed for disinfecting industrial water. Dependent on user requirements, the systems can be supplied in an inexpensive polyvinyl chloride (PVC) version or in stainless steel. Due to the availability of numerous sizes, it is possible to vary flow requirements—from 0.8 to 10 m^3/h. The cost-efficient PVC plastic UV system shown in Figure 3.7 was designed for the purification

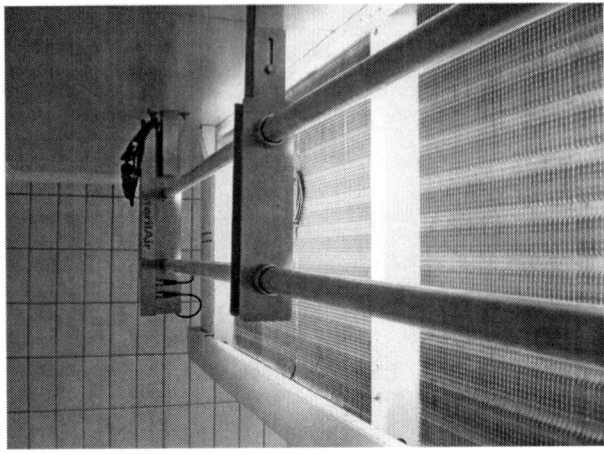

Figure 3.8 Ceiling-mounted UV disinfection system from sterilAir. (Credit to sterilAir, Weinfelden, Switzerland.)

of drains and process water. Measured at a default UV light dose of 400 J/m^2, its disinfection efficiency depending on the type is between 650 and 2800 L/h. Servicing and maintenance routines, such as the replacing of tubes, can be carried out without the use of tools.

3.3 DISINFECTION OF NONFOOD CONTACT SURFACES

Mold and biofilms can develop on nonfood contact surfaces such as ceilings, walls, floors, and equipment including tanks and vats, cooling coils, and food contact surfaces of equipment such as slicing and cutting units and conveyor belts (Kowalski, 2006). In general, standard cleaning and disinfection procedures are adequate to contain these problems but alternatives are available, including antimicrobial coatings like copper and TiO_2. UV irradiation of food processing equipment and surfaces, cooling coils disinfection systems, whole-area UV disinfection, and after-hours irradiation of rooms when personnel are not present are all viable control options for maintaining high levels of sanitation and disinfection in food processing facilities. These structure systems are built more as traditional type UV systems for surface disinfection. The UVC radiation is bundled with the assistance of special reflectors and then focused on the surface to be radiated. The example of UV disinfection system that can be mounted on the ceiling of the processing facility is shown in Figure 3.8.

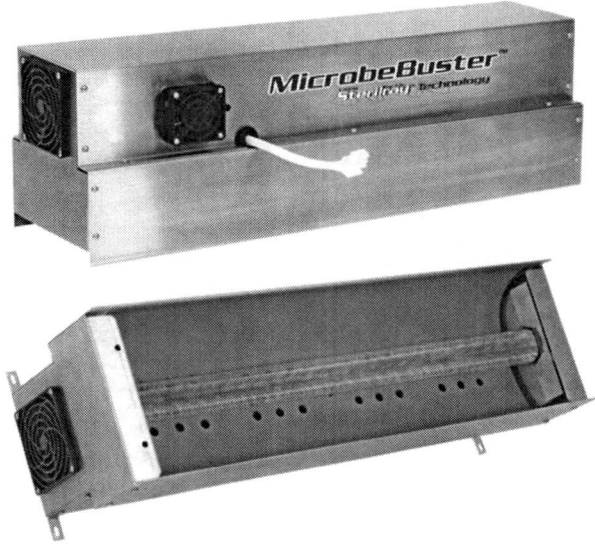

Figure 3.9 Sterilay MicrobeBusters for UV surface treatment. (Credit to HEI, Dover, NH.)

3.4 DISINFECTION OF FOOD CONTACT SURFACES

In food processing, the term "surface" includes the surface of the product, package, transport of product, and tools of the equipment that touch the product. The quality control of the surface based on application of UV light can be applied in the following locations: surface of incoming packaging, transport of product (i.e., belt, tray), post package—through package treatment and surface of the tools. When using UV light for surface treatments, the application and environment become as critical as the product characteristics.

The example of multiple purpose commercial surface application is MicrobeBusters UV units (Figure 3.9) from HEI that are designed to disinfect surfaces of conveyors, packages, bottle caps, etc. at all critical points along a production line. These units can be used in series or parallel to achieve the desired food safety objective (FSO) using excimer lamps in Far UV range.

Premium UV Modules (Figure 3.10) have been developed by Heraeus Noblelight GmbH (Hanau, Germany) especially for the food industry. Disinfection modules can be installed in filling and closing machines for dairy products and beverages and used to disinfect conveyor belts, transport containers, and work surfaces and the surfaces

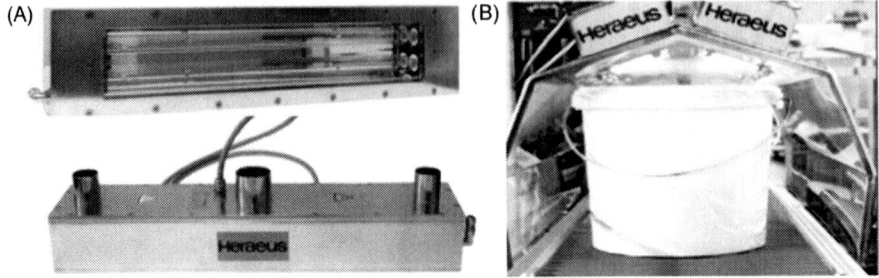

Figure 3.10 Premium UV module for surface disinfection (A) and example of a setup for food container disinfection (B). (Credit to Heraeus Noblelight GmbH, Hanau, Germany.)

of food. The system achieves a disinfection rate in excess of 99.9% with typical working distances of 20 mm and exposure times of 2–10 s. The packaging materials treated by the system achieve a degree of purity that corresponds to the classification as "low-germ packaging materials." Premium UV Module UV disinfection modules are ready-to-operate assembly systems consisting of one or more UV cassettes, LPA UV sources, fan housing, forced cooling with ambient air and power supply. UV irradiance at the beginning of lifetime at a distance of 20 mm is 65 mW/cm^2, at a distance of 100 mm is 45 mW/cm^2, and at a 200 mm is 26 mW/cm^2. Premium UV Module meets IP-68 classification and thus is suitable for CIP wash-down environments.

3.4.1 Conveyor Belts

Transportation belts are one of the most important prerequisites in the food industry to maintain hygiene and quality of products. UV light has a potential to kill up to 99.9% of total bacteria on conveyor belts for transporting food products and can be applied for both disinfecting surfaces of the product on the conveyor belts during transportation and the surface of conveyor belts. Several types of commercial equipment with UV light-assisted conveyor belt treatment are available for food processors. This equipment can be an easy and affordable solution to control the growth of psychotropic bacteria in finished products area.

The modular conveyor belt disinfection unit (Figure 3.11, sterilAir, Switzerland) can serve the demand in a flexible base frame installation. The disinfection system allows the decontamination of belts with a width of up to 1.50 m. A tilt switch can be fitted optionally. The unit is available in various lengths and power classes. The UV tubes are shatter protected and therefore comply with HACCP guidelines.

UV Disinfection of Air, Water, and Surfaces 25

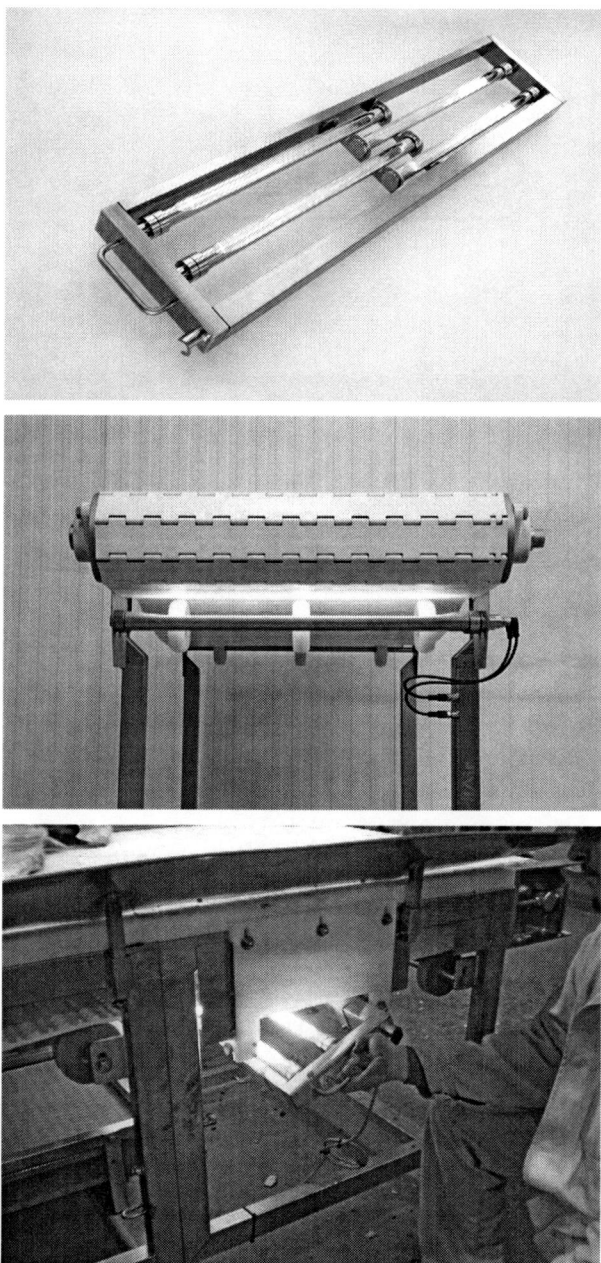

Figure 3.11 The sterilAir modular conveyor belt disinfection unit and the example of installation in meat processing facility. (Credit to sterilAir, Weinfelden, Switzerland.)

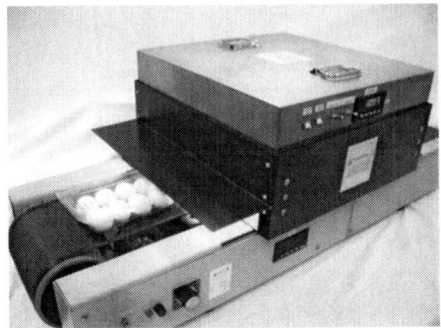 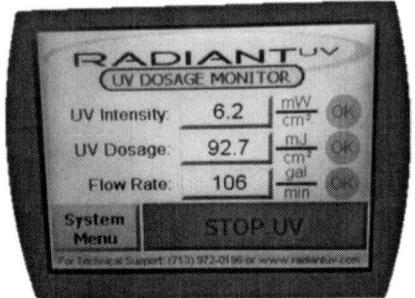

Figure 3.12 Conveyor UV disinfection system and UV dosage monitor from Radiant UV. (Credit to Radiant UV, Houston, TX.)

Direct UV irradiation of food is prohibited in the European Union. The surface disinfection has to be done on the returning belt. This makes it much more important to get the UV installation water, dirt, and shatter protected.

Radiant UV (Houston, TX) offers custom-designed UV light equipment for specific control needs of the application. The unit can be adjusted to fit not only the required decontamination level of the product but its quality and sensory characteristics. The UV unit from Radiant UV shown in Figure 3.12 can be used for packages disinfection prior to product packaging, product to package loading, and through package final disinfection. The systems are equipped with measurement system to control applied and delivered UV dose that displays UV power (ON & OFF), UV lamps intensity, and time of operation. Metering allows UV treatment serve as a critical control point (CCP).

Fuller UV, Inc., Frankfort, IL (IL, USA) offers a simple adjustable stand with fan-cooled UVC lamps that can be installed over conveyor belt and to disinfect surfaces of conveyor and packaged finished products. Variable conveyor speed and UV lamps height allow adjusting total applied UV dose for a variety of product sizes and shapes. More sophisticated solution for UV disinfection of conveyors belts that is based on SteriBelt high power disinfection modules can be found from Peschl UV Consulting (www.uv-conculting.de). The LPA lamps used in SteriBelt are characterized by higher UVC output power for the same lamp length. Moreover, because of the unique flat lamp technology, more than 50% of the generated UVC power is emitted directly onto the target surface. Other features include completely water-tight design, lamp shatter protection, simple, fast lamp change-over through preassembled plug connections, robust, compact stainless steel housing with no external air- or water cooling, and optimized for low ambient temperatures.

Other recommended areas of application of SteriBelt modules can include disinfection of conveyor belts in meat processing industry, as well as disinfection of sealing caps in filling machines and packaging in restricted areas.

3.4.2 Packaging

The packaging technologies play important role in extending the shelf life and protecting finished food products from postcontamination during storage. UV light can be applied as pre- or postpackaging technology to address issues associated with microbial spoilage. As a prepackaging control measure, UV treatment of packaging in filling plant, e.g., lids, cups, sealing, and packaging foils for drinks and beverages, helps to eliminate microbial contamination and thus extend shelf life of products. It was found that Gram-negative microorganisms are more UV sensitive than Gram-positive on the plastic surfaces. Spore forms such as *Bacillus* and *Aspergillus niger* are far more UV resistant than vegetative forms.

Claranor (Avignon, France) offers systems using PL to decontaminate tubs and cups in filling lines that handle sensitive products (Figure 3.13). The Claranor's installation consists of the electronics bay powered by the main current that generates electrical pulses and the optical cavity that focuses PL toward the surface that needs to be treated. The integrated cooling system regulates the temperature of the water in the lamp circuit. According to manufacturer, the PL technology guarantees a homogeneous treatment and is compatible with high output rate of 60 strokes/min. The Claranor's system (http://claranor.com/products/caps-sterilization2) can treat a wide range of types and sizes of cups; preformed and form fill seal, up to 15 cm depth. It is perfectly adapted for thermally treated products stored in cold chain for cups sterilization of fruit-based desserts, cups disinfection of dairy products, juices, ready meals filled at cold temperature. The application of PL for packaging sterilization for thermally treated products packaged in aseptic conditions such as soups, sauces, creams, fruit preparations, and stewed fruits is currently under development.

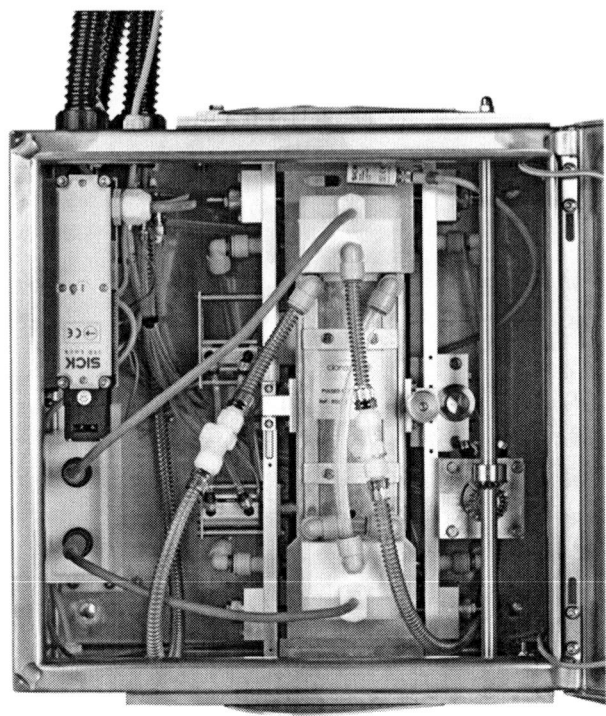

Figure 3.13 Claranor PL unit for closure sterilization. (Credit to Claranor PL, Avignon, France; http://claranor.com/products/caps-sterilization2.)

Figure 3.14 Radiant UV disinfection unit for packaging. (Credit to Radiant UV, Houston, TX.)

Radiant UV (Houston, TX) introduced UVC disinfection at 254 nm as a control method for package protection (Figure 3.14). The unit designed to deliver to 6 log reduction of *E. coli* and equipped with remote

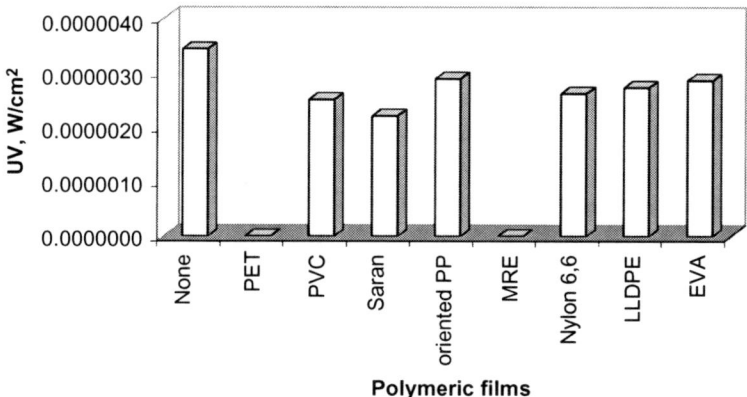

Figure 3.15 Comparison of UV transparency of selected polymeric films.

monitor measurement that allows to control UV dose, UV intensity, and exposure time.

When continuous UV and PL are applied as postpackaging treatment for finished packaged products, the considerations about transparency are referred to the packaging materials. For example, materials such as glass, polystyrene, and polyethylene (PET), which allow visible light to penetrate through the container, are not transparent to the UV light wavelengths that are essential for microbial inactivation and therefore they are not suitable for UV and PL treatments. Figure 3.15 shows the comparison of UV light permeability of eight polymeric films—PET, PVC, Saran, oriented PP, laminate material for Meals-Ready-to-Eat (MRE) pouches, nylon, low-density polyethylene (LDPE), and Ethylene vinyl acetate (EVA). PET and laminated films that are used to make pouches for MRE are not UV transparent due to the protective aluminum layer. On the other hand, polymers such as saran, polypropylene, PVC, EVA, nylon, and LDPE can transmit portions of UV light and hence meet the requirements for PL and UV treatments.

In addition, ink printed labels or drawings could interfere with the light absorption of the treated item and should be avoided on the surface of packaging materials. Besides the intrinsic transparency of the material, it is critical that the "condition" of the item to be treated is suitable for the penetration of the light. This means that the product surface should be smooth, clear and without roughness, pores and grooves which could "shadow" the microbial cells from the light, causing less complete light diffusion and thus reducing process effectiveness;

for the same reason, the item to be treated should be clean and free of contaminating particulates. In addition, items having a complex geometry could have areas hidden from the light and could require a more accurate design of the treatment chamber in order for the light to reach each point of the product surface. Another application of UV and PKL is a final intervention step for decontamination of the surface of the packaged food in the clean production areas that will be placed directly into the package with ready-to-eat (RTE) products such as packets of sauces, jams, and other items. The study of Keklik et al. (2009) demonstrated that PL UV light was effective against *Listeria monocytogenes* Scott A on the surfaces of unpackaged and vacuum-packaged (with polypropylene) chicken frankfurters. The optimum treatment conditions for both unpackaged and vacuum-packaged chicken frankfurters were found to be at 8 cm for 60 s, which resulted in about 1.6 log and 1.5 log CFU/cm^2 reductions (both approximately 97%) on unpackaged and vacuum-packaged samples, respectively.

3.4.3 Slicing Knives

Slicing knives can facilitate products cross-contamination without proper cleaning and sanitation of their surfaces where bacterial cells may be habituated. The applicability of the intense light pulses (PL) for decontamination of a stainless steel meat contact surface, exemplified by a slicing knife, was reported by Rajkovic et al. (2010), as a function of time between contamination and decontamination, number of light pulses applied, and the prior contact with different meat matrices. The reported results demonstrated successful application of PL treatment for reduction of *L. monocytogenes* and *E. coli* O157:H7 on a surface of stainless steel slicing knife. The inactivation effectiveness is depended on the type of meat product that was in the contact with the treated surface and on the time between the contamination and the PL treatment. The complete inactivation of 6.5 log CFU/side of knife was obtained when the knife surface was in contact with the products containing lower fat and protein content and when it was treated with PL as fast as possible after the contamination (within 60 s). Multiple light pulses due to the extended time between the moment of contamination and PL treatment could not improve the decontamination efficacy of PL treatment. Results showed that the suggested approach could be very effective as an intervention strategy in meat processing lines for preventing cross-contamination between the slicing equipment and the final product.

CHAPTER 4

Case Studies of UV Treatment of Food Surfaces

Comprehensive reviews in the area of UV and PL applications for food surfaces have been compiled by the US FDA (2000b) and Woodling and Moraru (2005). The variability of the results (a 2- to 8 log reduction was generally reported) is most likely due to the different challenge microorganisms used in various studies, the light intensity of the treatment, and the different properties of the treated substrates. Woodling and Moraru (2005) demonstrated that the efficacy of PL is affected by substrate properties such as topography and hydrophobicity, which affect both the distribution of microbial cells on the substrate surface and the interaction between light and the substrate (i.e., reflection and absorption of light). In designing a PL treatment for food items, both source (as light wavelength, energy density, duration and number of the pulses, interval between pulses) and target food (as product transparency, color, size, smoothness, and cleanliness of surface) parameters are critical for process optimization, in order to maximize the effectiveness of product microbial inactivation and to minimize the product alteration. Such alteration can be mainly determined by an excessive increase of temperature causing thermal damage to foods but also by an excessive content of UVC light which could result in some undesired photochemical damage in food itself or packaging materials.

4.1 FRESH PRODUCE

Recent studies of the germicidal effects of UV light against naturally occurring pathogenic and nonpathogenic microflora on the surface of fresh produce can be synergistically enhanced by the hormetic response of irradiated fruits. For instance, Li et al. (2010) reported higher inhibition of *Monilinia fructicola* growth in the pears inoculated with the pathogen before the UVC treatment than in those being inoculated after UVC exposure. Similarly, Pombo et al. (2011) observed reduction in growth of *Botrytis cinerea* inoculated on the strawberries 8 h after UVC treatment (4.1 kJ/m^2). In other studies, Obande et al. (2011) studied the shelf life of tomatoes that was first exposed to UVC light at 8 kJ/m^2 and then was inoculated with *Penicillium digitatum*. After 10 days of storage at 20°C, the UV-treated fruits were firmer and the diameter of fungal lesion was considerably smaller in comparison to controls. Therefore, higher resistance to postharvest diseases of UV-treated commodities can be partially attributed to the physiological changes stimulated by UV light.

Besides the molds, on the surface of fresh produce can be present pathogenic bacteria, such as *Salmonella* spp., O157:H7 and non-O157 shiga toxin producing *E. coli* that constitute a threat to human health and safety. It was presented by several authors that either UVC or PL treatments have ability to reduce the population of these pathogens. For instance, Yaun et al. (2004) reported reduction of *E. coli* O157:H7 by approximately 3.3 logs on the apples exposed to UVC light at 240 W/m^2. Same UV irradiation conditions resulted in slightly lower log reduction of *Salmonella* spp. on tomatoes (2.19 logs). PL (Xenon Corp.), with the emission spectrum in the UV/visible range (100–1100 nm), was applied for 5, 10, 30, 45, and 60 s to raspberries inoculated with *E. coli* O157:H7 and *Salmonella* spp. Bialka et al. (2008) reported reductions between 0.7 and 3.0 log$_{10}$ CFU/g of *E. coli* O157:H7 and 1.2 and 3.4 log$_{10}$ CFU/g of *Salmonella* on treated berries. However, fruit processing with PL was accompanied by the temperature increase, and therefore microbial reduction might result from the combined light-heat effects. These examples demonstrated that the postharvest UV processing of variety of fresh produce could be effective against both pathogenic and nonpathogenic microflora. More cases of successful UV applications are presented in Table 4.1.

Table 4.1 UV Treatments of the Surface of Fresh Fruit Commodities			
Commodity	UV Treatment Lamp/ Number/Output Power/ Fluence	Germicidal Effects	References
Apples	UVC/1/30 W/7.5 kJ/m^2	Enhanced resistance against alternaria rot, brown rot (*Monilina* spp.), and bacterial soft rot (*Erwinia* spp.)	Lu et al. (1991)
	UVC/1/NA/240 kJ/m^2	3.3 log$_{10}$ reduction of *E. coli* O157:H7	Yaun et al. (2004)
Blueberries	Pulsed UV/visible light/ 60 s (22.6 J/cm^2)	4.3 log$_{10}$ reduction of *E. coli* O157:H7; 2.9 log$_{10}$ reduction of *Salmonella* spp.	Bialka and Demirci (2007)
Raspberries	Pulsed UV/visible light/ 60 s (59.4 J/cm^2)	3.0 log$_{10}$ reduction of *E. coli* O157:H7; 3.4 log$_{10}$ reduction of *Salmonella* spp.	Bialka et al. (2008)
Strawberries	Pulsed UV/visible light/ 60 s (59.4 J/cm^2)	2.3 log$_{10}$ reduction of *E. coli* O157:H7; 3.9 log$_{10}$ reduction of *Salmonella* spp.	Bialka et al. (2008)
Tomatoes	UVC/1/NA/240 W/m^2	2.19 log$_{10}$ reduction of *Salmonella* spp.	Yaun et al. (2004)

4.2 FRESH-CUT PRODUCE

Fresh-cut fruits become more popular among consumers due to increased preference for minimally processed fresh-like and RTE products. Mechanical operations of fresh-cut fruits production, such as peeling, slicing, shredding, etc., often result in enzymatic browning, off-flavors, texture breakdown, and lower resistance of fresh-cut produce to microbial spoilage in comparison with the unprocessed commodities because of the presence of natural microflora on the surface of raw commodities. Therefore, during operations of cutting and shredding, the cross-contamination may occur that might increase the risks of foodborne outbreaks.

To improve the hygiene and safety during the mechanical processing, sanitizing and dripping treatments are commonly applied. During washing and dipping steps, raw or fresh-cut material is immersed into the tap water containing sanitizing agents (chlorine, sodium hypochlorite) to remove microorganisms, pesticide residues, and plant debris from product surface. To reduce the usage of sanitizing chemicals, UV light alone or in combination with ozone or another preservative agent was explored as novel processing alternatives. Fonseca and Rushing (2006) examined the effects of UVC light ($1.4-13.7$ kJ/m^2 at 253.7 nm) on the quality of fresh-cut watermelon compared to the common sanitizing solutions. Dipping cubes in chlorine (40 µL/L) and ozone (0.4 µL/L) was not effective in reducing microbial populations, and cubes quality was lower after these aqueous treatments compared to UV-irradiated cubes or control. In commercial trials, exposure of packaged watermelons cubes to UVC at 4.1 kJ/m^2 produced more than 1 log reduction in microbial populations by the end of the product's shelf life without affecting juice leakage, color, and overall visual quality. Higher UV doses did not effect microbial populations or result in quality deterioration (13.7 kJ/m^2). Spray applications of hydrogen peroxide (2%) and chlorine (40 µL/L) without subsequent removal of excess water failed to further decrease microbial load of cubes exposed to UVC light at 4.1 kJ/m^2. It was concluded that when properly utilized, UVC light is the only method tested that could be potentially used for sanitizing fresh-cut watermelon. Similarly, exposure of sliced apples to UVC resulted in higher (~1 log) reduction of *L. innocua* ATCC 33090, *E. coli* ATCC 11229 and *S. cerevisiae* KE 162 in comparison to the apples pretreated with antibrowning and sanitizing agent (1% w/v

Table 4.2 Effect of UVC Light on Reduction of Microorganisms in Fresh-Cut Produce

Fresh-Cut Commodity	Microbiological Organism	Number/UV Lamp/Power/ Fluence	References
Watermelon	Mesophilic, psycrophilic, and enterobacteria	15/LPM/36 W 1.6, 2.8, 4.8, and 7.2 kJ/m^2	Artés-Hernández et al. (2010)
Cantaloupe Melon	Yeast, mold, *Pseudomonas* spp., mesophilic aerobes, lactic acid bacteria	1/LPM/N/A 0.0118 kJ/m^2	Lamikanra et al. (2005)
Apple	*L. innocua* ATCC 33090; *E. coli* ATCC 11229; and *S. cerevisiae* KE 162	2/LPM/15 W 5.6 ± 0.3, 8.4 ± 0.5, and 14.1 ± 0.9 kJ/m^2	Gomez et al. (2010)
Pears	*L. innocua* ATCC 33090; *L. monocytogenes* ATCC 19114 D; *E. coli* ATCC 11229; and *Zygosaccharomyces bailli* NRRL 7256	2/LPM/15 W 15, 31, 35, 44, 56, 66, 79, and 87 kJ/m^2	Schenk et al. (2007)

ascorbic acid—0.1% w/v calcium chloride). The combination of UVC with antibrowning pretreatment better preserved color of sliced apples during storage at 5°C for 7 days (Gomez et al., 2010). Other studies have shown that UVC treatment applied alone was efficient in reduction of the number of microbiological organisms present on the surface of fresh-cut crops. The examples of successful applications of UVC light are given in Table 4.2.

Similarly to raw crops, the effectiveness of UV treatment on reduction of microbial deterioration and quality retention was defined by the delivered UV dose and overall characteristics of the surface exposed to the UV light. Lamikanra et al. (2005) stressed out that moment of the operational step of UV light during the fruit processing is an important factor. In their studies, the authors exposed the cantaloupe melon to UVC at 254 nm during cutting and after cut of the fruits. Cutting of cantaloupe melon under the UVC light was as effective as postcut treatment in reduction of yeast, molds, and *Pseudomonas* spp. populations. However, fruit cutting during simultaneous exposure to UVC resulted in improved product quality, i.e., reduced rancidity and respiration rate, and also increased firmness retention, when compared to postcut and control samples. Better preservation of fruits processed during the UV exposure can be related to the defense response of the wounded plant enhanced by the UV. Mechanical injury of the plant tissues activates the expression of wound-inducible genes. UV radiation is capable to induce the expression of plant defense-related proteins that are normally activated during wounding. For example, Lamikanra et al. (2005) reported significant increase in ascorbate peroxidase enzyme activity during storage of cantaloupe melon processed under UVC light. Peroxidases protect plant cells against the oxidation. In terms of UV effects on fruits flavor, Lamikanra et al. (2005) reported that fruits processed with the UV light preserved their aroma to the same extent as nontreated control samples.

4.3 UV- AND PL-BASED PROCESSING AIDS FOR MEATS

Among known applications of UV light in meat processing are UV treatments of nonfood and food contact surfaces and recently, product surfaces. UV light was effective against *E. coli* and *Salmonella* on pork skin, raw meats, and poultry surfaces. The precautions should be taken to avoid undesirable brown discoloration of red meat since UV encourages metmyoglobin formation. The rate of product discoloration depends on the intensity of light and wavelength distribution. Packaged product can be exposed to UV so to avoid of risk of recontamination. The correct choice of polychromatic pulsed or monochromatic continuous source in combination with the UV filter and light permeability of the packaging material is critical when such a process is developed for red meat products.

According to USDA, contamination of foods with *L. monocytogenes* is likely to occur in all RTE meat and poultry products that are exposed to postlethality processing environment. These postlethality processing environments are the areas into which products are routed after complete treatments and may include slicing, peeling, dicing, rebagging, and brining. As an example of UV technology postlethality application, packaged RTE product can be exposed to UV so to avoid of risk of cross or recontamination. Knowledge of light permeability of the packaging material is critical when such a process is developed.

USDA research (Sommers et al., 2009) has demonstrated that UV light can be used to reduce levels of *Listeria* contamination on the surfaces of RTE products. UVC irradiation of frankfurters that were surface inoculated with *L. monocytogenes* resulted in a 1.31, 1.49, and 1.93 log reduction at doses of 1, 2, and 4 J/cm^2, respectively. UVC treatment had no effect on frankfurter color or texture at UVC doses up to 4 J/cm^2. It was concluded that because the numbers of *L. monocytogenes* associated with contaminations of RTE meats are typically very low, the use of UVC in combination with potassium lactate and sodium diacetate has the potential to reduce the number of frankfurter recalls and foodborne illness outbreaks (Sommers et al., 2010).

In recent studies, UV light was reported to be effective against nonspore forming pathogenic microflora on raw meats and poultry surfaces. UVC at $0.5\,\text{J/cm}^2$ reduced the initial populations of *Campylobacter jejuni, L. monocytogenes,* and *Salmonella typhimurium* by 1.3 to 1.2 log CFU/g, respectively on chicken breasts (Chun et al., 2010). Similarly, UV treatment of raw chicken fillet at $0.2\,\text{J/cm}^2$ reduced *C. jejuni, E. coli*, serovar *Enteritidis*, total viable counts, and Enterobacteriaceae by 0.76, 0.98, 1.34, 1.76, and 1.29 log CFU/g, respectively. Following UV treatment of packaging and surface materials, higher reductions of up to 3.97, 4.50, and 4.20 log CFU/cm^2 were obtained for *C. jejuni, E. coli*, and serovar *Enteritidis*, respectively. These studies indicate that UV may be beneficial in working areas of meat and poultry processing plants to reduce the level of aerobic and nonspore forming pathogens and may be applied to decontaminate cut-up products moving on conveyors, associated packaging, and surface materials. In order to demonstrate efficacy of UV technology to USDA-FSIS, the UV dose requirements to achieve specific log reduction requirements need to be evaluated first. This information will allow selecting not only a correct UV source but also effective design of a system depending on the application.

Meat processors may dedicate a full shift to cleaning disassembled equipment, conveyor belts, walls, ceilings, and floors because *Listeria* can survive for extended periods in meat processing facilities. The benefits of UV units installed in cold rooms are not limited to killing bacteria on surfaces and in the air. A substantial extension of storage life of chilled carcasses is possible when UV light is used continuously and delays the onset of microbiological spoilage. Additionally, UV reduces shrinkage and retains juices, cold rooms' odors are eliminated, and mold growth on the surfaces is reduced.

4.4 SHELL EGGS

The majority of *Salmonella enteritidis* outbreaks has been related to the consumption of raw or undercooked eggs or egg-containing foods. Therefore, the USDA mandates egg washing for all graded eggs by use of a detergent solution and sanitizer. These agencies and the egg industry have been investigating alternative decontamination techniques, which could better serve the public, minimize costs, and benefit both the public and the industry. Keklik et al. (2010) studied the effectiveness of PL UV light for the decontamination of eggshells. Eggs inoculated with *S. enteritidis* on the top surface at the equator were treated with pulsed UV. A maximum log reduction of 5.3 CFU/cm^2 was obtained after a 20-s treatment at 9.5 cm below the UV lamp at a total dose of 23.6 ± 0.1 J/cm^2, without any visual damage to the egg. cUV light has been documented to be effective in reducing various bacterial populations on egg shell surfaces including total aerobic plate count (Chavez et al., 2002), *S. typhimurium* and *E. coli* (Coufal et al., 2003), and *Yersinia enterocolitica* (Favier et al., 2001). Kuo et al. (1997) reported UV inactivation of aerobic bacteria and molds in addition to *S. typhimurium*. Ninety nine percentage reduction of CFU of aerobic bacteria per egg was observed for all UV treatments during 0, 15, and 30 min at the intensity of 0.62 mJ/cm^2. Despite the urgent need to improve egg safety and demonstrated UV light inactivation efficiency, UV treatment of eggs has not yet been commercially implemented.

4.5 SEAFOOD

The outbreaks of fish and fish products are on the top of the list of foods associated with foodborne diseases. To enhance shelf life and enhance safety of seafood products, UV light was studied in terms of microbial inactivation efficiency.

Huang and Toledo (1982) found that UV light at the doses of 300 mWs/cm^2 from a high-intensity UVC lamp reduced surface microbial count on mackerel by 2–3 log cycles. UV-treated mackerel wrapped in 1 mil polyethylene and packed in −1°C ice had at least a 7-day longer shelf life than conventional ice-packed untreated controls. Spray washing with water containing 10 ppm chlorine by itself or in combination with UV light was necessary to reduce surface counts on rough-surfaced fish to the same extent as that on smooth-surfaced fish. Inactivation of *E. coli* O157:H7 and *L. monocytogenes* inoculated on raw salmon fillets by PL UV-light treatment was investigated by Ozer and Demirci (2006). The PL system from xenon generated 5.6 J/cm^2 per pulse at the lamp surface and three pulses per second. In order to avoid overheating, the fillets were exposed to pulses at the distance of

8 cm, and at least 1 $\log_{10}$ CFU reduction of *E. coli* and *L. monocytogenes* was achieved after 1-min treatment.

4.6 UV EQUIPMENT FOR TREATMENT OF FOOD SURFACES

An example of a commercially available system to decontaminate surfaces of foods is the UV tumbling process that is offered by Reyco Systems (USA) (http://www.reycosystems.com/solutions/uv-drum/). The company incorporated either a rotating drum or a screw conveyor that lifts and tumbles the product to ensure exposure to the UV source. The unit can be used to treat fresh products including meats, etc., frozen products (meats, seafood), and cooked, refrigerated products. The company used the patented technology of Steril-Aire™ UV Emitters that are sleeved in plastic to meet the food safety requirements of food processing facilities. The patented design allows emitters to work efficiently in the cold environment of refrigerated or chiller coils, where competitive units lose their effectiveness. Examples of UV units currently used in commercial processing facilities include: (i) UV tumbling drum in operation for chicken and beef fajita strips (cooked and Individually Quick Frozen (IQF) frozen) with a capacity 6000–7000 lb/h; (ii) cooked and IQF frozen hamburger patty treatment (hooded conveyor with turn over), with a capacity 3000 lb/h; and (iii) a deli meat system (custom conveyor with UVC hood) for formed deli ham logs, with a capacity of 10,000 lb/h.

Pathogen reduction box from HEI (Dover, NH) was used to test the effect of far UV range at the storage of whole tomatoes and potatoes in the container at ambient room temperature. The exposure of the tomato fruits to UV light at the dose of 1000 mJ/cm^2 extended their shelf life up to 14 days as shown in Figure 4.1A. Potatoes exposed to UV light in similar conditions did not show signs of the mold growth on the surface during more than 57 days as shown in Figure 4.1B.

Case Studies of UV Treatment of Food Surfaces 39

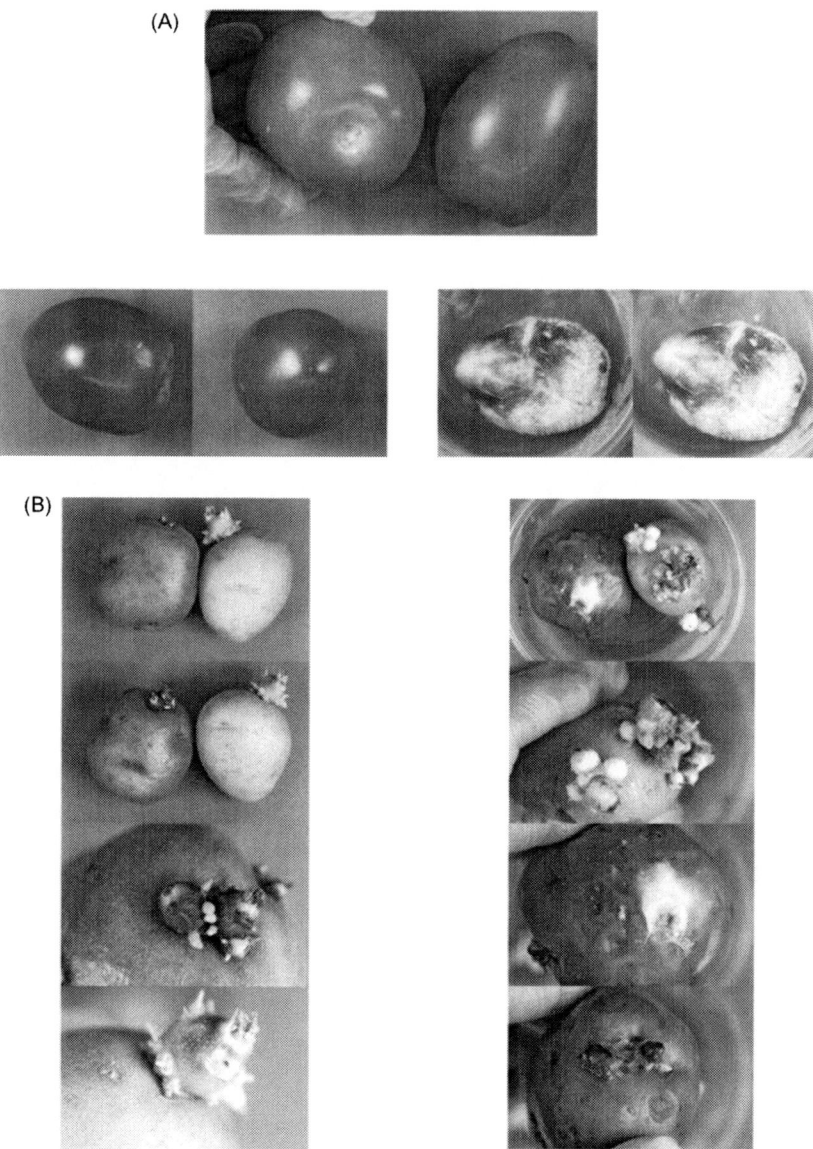

Figure 4.1 (A) Comparison of shelf life of untreated tomatoes and tomatoes after exposure to UV light in the pathogen reduction box. (B) Comparison of shelf life of untreated potatoes and potatoes after exposure to UV light in the pathogen reduction box. (Credit to HEI, Dover, NH.)

CHAPTER 5

Principles of UV Surface Process Development

5.1 FACTORS AFFECTING INTERACTION BETWEEN UV LIGHT AND SURFACE OF MATERIALS

UV light emitted from the gas discharge of a UV source propagates away from atoms and ions interacting with the materials through absorption, reflection, refraction, and scattering. This explains why correct positioning of UV source and distance from treated product is critical to maximize efficacy of UV treatments. Any obstruction to the path of the light, such as dust, shadowing or clumping of bacteria, can reduce efficacy. So, the effectiveness of UV light is less on a rough surface than on a smooth one.

The characteristics of UV source such as wavelength or the levels of UV light photons energy, the number of photons or UV intensity, and exposure time will define UV process lethality. It was found that among product parameters temperature, chemical composition, and physical structure affect UV light effectiveness to control microorganisms. Additionally, the efficacy of UV sources against different groups of pathogenic and a spoilage organism varies in different matrices such as air, water, surfaces, or fluid foods. Adenovirus is considered as the most UV-resistant organism in water treatment, whereas the highest UV resistance in air and food matrices characterizes bacterial and molds spores.

5.2 DETERMINATION OF UV DOSE

The required dose (UV req) or the UV dose in units of millijoule per square centimeter that is needed to achieve the target logarithmic inactivation or specific logarithmic reduction (SLR) on the surface of material for the target pathogen dose can be determined from Eq. (5.1) as a product of D_{UV}-value and SLR.

$$\text{UV req} = \text{SLR} \times D_{UV}\text{-value} \qquad (5.1)$$

To calculate the required UV dose, the decimal reduction dose (D_{UV}-value) or UV dose to achieve 1 log reduction of target pathogen has to be known. The data in the scientific literature reported that there is a difference in UV sensitivity of bacteria depending on the surface characteristics. For example, D_{UV}-values of *Listeria*, summarized in Table 5.1, have been evaluated based on the inactivation data on the agar surfaces, conveyor belts, and products such as pear, frankfurters, and ham. It can be seen from Table 5.1, that UV light resistance of *L. monocytogenes* depends on the nature of surfaces and its roughness characteristics. For a smooth surface, the decimal UV dose varies from 2.5 up to 3.5 mJ/cm² whereas for the product surface the UV dose may be higher by 2 or 3 orders of magnitude and vary in the range of 200–300 mJ/cm².

Applied UV fluence or applied dose in millijoule per square centimeter that is generated by an applied incident UV light emitted by the

Table 5.1 UV Resistance of *Listeria* Strains on Food Contact and Food Surfaces

Listeria Strains	Medium	UV Intensity and Treatment Time, mW/cm²	Reported Log Reduction	Estimated D_{UV}-value, mJ/cm²	Source
Serotypes 3A, 4A, 4B, and 4C	Conveyor belts 1. Ropanyl DM 8/2 2. Volta FRMW-3.0 3. Volta FRMB-3.0 4. Ropanyl DM	5.95 mW/cm² for 3 s 5.53 mW/cm² for 3 s	7–5.27	2.55–3.19	Morey et al. (2010)
L. monocytogenes strains from RTE (F4561, H7762, and 121 H7764)	Frankfurters	5 mW/s/cm² for 100 s	2.32–2.80	215–178	Sommers et al. (2009)
L. innocua	Frankfurters with sodium diacetate and potassium lactate (SDA/PL)	1000 mJ/cm² followed by flash steam (0.75 s steam/121°C) 10 mW/cm² for 100 s	3.19	300	Sommers et al. (2010)
L. monocytogenes ATCC 1911	Agar surfaces Sliced ham	0.1–1.8 mJ/cm² 800 mJ/cm²	0.9–3.9 2.57	0.48 311	Chun et al. (2009)
L. monocytogenes ATCC 1911	Cut pear	2 LPM, 15 W 1500–8700 mJ/cm²	Up to 3.5	2500	Schenk (2007)

source on the surface of sample in a certain exposure time can be calculated from Eq. (5.2).

$$UV_{dose\ applied} = I_{UV} \times t \qquad (5.2)$$

where I_{UV} is the UV light intensity on the surface of the product that is a function of UV lamp output power and distance (mW/cm^2) and t is the exposure time (s). UV intensity is the power incident on a unit area perpendicular to the direction of propagation. UV intensity describes the magnitude of UV light measured with a radiometer. Applied dose or fluence reflects the energy emission from the UV source and it is independent to the material to be irradiated. Knowledge of the applied dose is important to select a correct power and type of UV source by taking into the account their UV efficiency. US FDA is limiting the level of UV light intensity on the surface that must be less than 215 or 0.215 mW/cm^2. In order to meet performance objectives, the actual UV applied dose has to be at least equal or higher than the required UV dose.

$$UV\ dose\ applied \geq UV\ required \qquad (5.3)$$

Knowledge of the applied dose allows evaluating the treatment time or selecting the UV bulbs that will suit food plant environment. The UV lamp type, output power, length, and the number of UV lamps will depend on the dimensions of the total area where product will be placed and treated. It should be considered, that LPM UV lamps convert approximately 38% of their input watts into UVC output watts. For example, if the required UVC is 84 W, the total lamp wattage is 84 W/0.38 = 221 W. The length of the UV lamp is a critical factor when establishing the dimensions of the UV exposure chamber. The lamp length should always be positioned between the inlet/outlet ports. To design or select commercially available UV system, the treatment surface area and the length of the lamp should be known.

5.3 ENVIRONMENTAL ASSESSMENT

Expected increase of world population up to 9 billion by 2050 brings the necessity to implement the sustainable practices that will allow meeting the needs of the present without compromising the ability of future generations to meet their own needs. These include wiser management of the natural resources use, product stewardship, strengthening energy efficiency, development of new technologies that reduce the consumption of resources and eradication of poverty. UV light is an emerging nonthermal technology that has much to offer for the sustainable development of society. Its application for the food processing is energy and cost efficient and also was proven to yield the fresh-like, safe, and high-nutritional value products. Moreover, UV light applied as a postharvest technology can significantly reduce the loss of fresh produce, which in the developed countries is of the order of 20% and as high as 50% in developing countries (Obande et al., 2011). It was shown by many researchers that UV technology might be used as alternate method to control postharvest diseases caused by fungi. This in turn may substantially reduce the usage of fungicides as well as other chemicals that pose serious health hazard and environmental risks. The major disadvantage of UV technology is the mercury content in UV sources. The potential mercury exposure due to lamp sleeve breakage is a health concern. Breakage of lamps can occur when lamps are in operation and during maintenance. The mercury contained within a UV lamp is isolated from exposure by the lamp envelope and surrounding lamp sleeve. For the mercury to be released, both the lamp and lamp sleeve must break. The mercury content in a single UV lamp used for water treatment typically ranges from 0.005 to 0.4 g (5–400 mg). LPM lamps have less mercury (5–50 mg/lamp) compared to low pressure high output (LPHO) (26–150 mg/lamp) and MPM lamps (200–400 mg/lamp). The EPA established a maximum contaminant level (MCL) for mercury at 0.002 mg/L. The EPA has found mercury to potentially cause kidney damage from short-term exposures at levels above 0.002 mg/L MCL. The concern over the impact of mercury release into the food plant environment stimulated the development and validation of mercury-free special technologies lamps and LEDs. Xenon flash lamps are also more environmental-friendly than continuous UV lamps because they do not use mercury.

In Europe, the so-called cell lamps are the first choice emitters for installations in food production facilities. The cell lamp easily can be integrated in installations and can be highly water protected. Cell UV lamps are also much more temperature stable in terms of UV output under cooling conditions. More technical details can be found at http://www.sterilair.com/en/company/competence/technology.html. In addition to the ability of UV systems to work at the low ambient air temperature of 3–5°C, they have to stand hand-held manual cleaning made by high-pressure spray nozzles and meet US FDA HACCP requirements. This means that UV sources have to be easy to maintain and the installation has to be as simple as possible. For this reason, the electronic ballasts are integrated into the appliance itself. The energy consumption of cell lamps is lower compared to standard emitters under the same cold air conditions, as there is no need to compensate the internal pressure drop by a higher current.

5.4 REGULATORY STATUS

A variety of UV sources are commercially available or currently under development that can be applied for specific purposes at the food plant whereas LPM lamps and xenon PL are currently the dominant sources of radiation for the processing and treatment of foods since they were approved by the US FDA CFR 21 179.39 (Table 5.2).

Code 21CFR179.41, issued by the FDA in 1996, Department of Health and Human Services, approves the use of pulsed UV light in the production, processing, and handling of food. PL may be safely used for the treatment of foods under the following conditions: (i) The radiation sources consist of xenon flash lamps designed to emit broadband radiation consisting of wavelengths covering the range of 200–1000 nm and operated so that the pulse duration is no longer than 2 ms; (ii) The treatment is used for surface microorganism control; (iii) Foods treated with PL shall receive the minimum treatment reasonably required to accomplish the intended technical effect; and (iv) The total cumulative treatment shall not exceed 12.0 (J/cm^2).

In assessing the safety of foods treated with all forms of radiation, the agency considers microbial efficacy, changes in chemical composition of the food that may be induced by the proposed treatment, including any potential changes in nutrient levels. The legal status of

Table 5.2 US FDA CFR 21 179.39—UV for Treatment and Processing Foods		
Radiated Food	**Limitations**	**Use**
Food and food products	Without ozone production: high-fat content food irradiated in vacuum or in an inert atmosphere; intensity of radiation, 1 W (of 2537 A radiation) per 5–10 ft.^{2}	Surface microorganism control

light treatments in other parts of the world has a different approach, because the legislation is not technology oriented but food and food ingredient oriented. Light technologies would fall in the scope of regulation on novel foods and novel food ingredients. In order to provide an appropriate definition for what constitutes a "novel food," first one would have to identify their global location or global location of interest, as some countries have adopted specific term(s) and definitions (and regulations), whereas others have not. The definitions of Novel Foods are available in six countries EU, the Great Britain, Canada, Australia, New Zealand, and China. There are no regulations or formal definitions for "novel food" in the United States. GRAS (generally recognized as safe) regulations can serve as some analogy of Novel Foods in the United States. Before novel process can be used and product can be sold, the thorough reviews and evaluations of safety have to be conducted by regulatory agencies. Potential microbiological, toxicological, or nutritional concerns that can result from novel processing or preparation techniques have to be assessed. The petitioner must provide sufficient, scientific, and statistically sound information regarding process validation and produced foods assessment to prove that process consistently produces the product meeting its predetermined specifications and quality.

CHAPTER 6

Conclusions

Foodborne and waterborne pathogens are the largest group of microorganisms that present health hazards in the food industry. Some of these organisms can become airborne during processing and settle on raw, semifinished or finished products thereby becoming amenable to control by UV and PL in the air and on the surfaces. Virtually every food processing facility can benefit from the use of light-based technologies to control microbial hazards through treatments of air, nonfood, food contact surfaces, processing water, ingredients, and surfaces of raw and finished products.

UV light-based technologies are emerging both as mild thermal and nonthermal techniques that have much to offer for the sustainable development of society. Their application for the food processing is energy and cost efficient and also was proven to result in products with higher safety and better quality characteristics when used as intervention step and shelf-life extension method for different groups of foods. However, despite a large number of benefits of UV light applications in food industry, energy and processing water saving opportunities, enhanced safety need to be carefully considered in each specific case for successful technology implementation and to assure positive benefits. LPM and amalgam sources are mainly adapted by the industry because they are readily available at comparatively low cost along with commercial systems that are developed and can be successfully integrated in food facilities at various points of production. Cell LPM lamps are used in special conditions in food facilities in Europe due to their high-performance characteristics at low temperature and water protection. Future tunable monochromatic (EL and LEDs) and polychromatic sources would allow users to adjust the wavelength to specifically match the needs of the application. In recent years, UV LEDs have been developed with the following many advantages: small size, long lifetime up to 10,000 h, energy-efficient, easy control of emission, and no production of mercury waste and glass. The commercial UV LED can emit in three diapasons of UVA, UVB, and UVC. The germicidal effects against several bacteria or fungi have been demonstrated and reported along with the first applications for disinfections of hospital rooms and water. Are LEDs the right new technological solutions with enormous potential for meat processing operations to fight with pathogens? The near future will show.

Innovative research of UV light for food plants applications has grown worldwide aiming to overcome challenges and improve efficacy of treatment that generated new knowledge in this area. Prerecorded online courses offered by Novel Food Sciences discuss the state of the art of UV light technology for foods and food surfaces can be found and downloaded at http://novel-food-sciences.com/iclasses/indexengineering.php. More research and end user education can ensure the effectiveness of UV light applications and accelerate the technology transfer for microbial inactivation in foods, stimulate the growing interest in the nonthermal technologies, and assist in the successful commercialization for food processing applications.

REFERENCES

Artés-Hernández, F., Robles, P.A., Gómez, P.A., Tomás-Callejas, A., Artés, F., 2010. Low UV-C illumination for keeping overall quality of fresh-cut watermelon. Postharvest Biol. Technol. 55, 114–120.

Bialka, K.L., Demirci, A., 2007. Decontamination of Escherichia coli O157:H7 and Salmonella enterica on blueberries using ozone and pulsed UV-light. J. Food Sci. 72, M391–M396.

Bialka, K.L., Demirci, A., Puri, V.M., 2008. Efficacy of pulsed UV-light for the decontamination of Escherichia coli O157:H7 and Salmonella spp. on raspberries and strawberries. J. Food Sci. 73, M201–M207.

Chavez, C., Knape, K., Coufal, C., Carey, J., 2002. Reduction of eggshell aerobic plate counts by ultraviolet irradiation. Poult. Sci. 81, 1132–1135.

Chun, H.H., Kim, J.Y., Song, K.B., 2009. Inactivation kinetics of Listeria monocytogenes, Salmonella enterica serovar Typhimurium, and Campylobacter jejuni in ready-to-eat sliced ham using UV-C irradiation. Meat Sci. 83 (4), 599–603.

Chun, H.H., Kim, J.Y., Lee, B.D., Yu, D.J., Song, K.B., 2010. Effect of UV-C irradiation on the inactivation of inoculated pathogens and quality of chicken breasts during storage. Food Control 1, 276–280.

Coufal, C.D., Chavez, C., Knape, K.D., Carey, J.B., 2003. Evaluation of ultraviolet light sanitation of broiler hatching eggs. Poult. Sci. 82, 754–759.

Favier, G., Escudero, M., De Guzman, A., 2001. Effect of chlorine, sodium chloride, trisodium phosphate and ultraviolet radiation on the reduction of Yersinia enterocolitica and mesophilic aerobic bacteria from eggshell surface. J. Food Prot. 64 (10), 1621–1623.

Fonseca, J.M., Rushing, J.W., 2006. Effect of ultraviolet-C light on quality and microbial population of fresh-cut watermelon. Postharvest Biol. Technol. 40, 256–261.

Gómez, P.L., Alzamora, S.M., Castro, M.A., Salvatori, D.M., 2010. Effect of ultraviolet-C light dose on quality of cut-apple: microorganism, color and compression behavior. J. Food Eng. 98, 60–70.

Gomez-Lopez, V.M., Ragaerta, P., Debeverea, J., Devlieghere, F., 2007. Pulsed light for food decontamination: a review. Trends Food Sci. Technol. 18, 464–473.

Groenewald, W.H., Gouws, P.A., Cilliers, F.P., Witthuhn, R.C., 2013. The use of ultraviolet radiation as a non- thermal treatment for the inactivation of Alicyclobacillus acidoterrestris spores in water, wash water from a fruit processing plant and grape juice concentrate. J. New Gen. Sci. 11 (2), 19–32.

Huang, Y.-W., Toledo, R., 1982. Effect of high doses of high and low intensity UV irradiation on surface microbiological counts and storage-life of fish. J. Food Sci. 47, 1667–1669.

Keklik, N.M., Demirci, A., Puri, V.M., 2009. Inactivation of Listeria monocytogenes on unpackaged and vacuum-packaged chicken frankfurters using pulsed UV-light. J. Food Sci. 74, 8.

Keklik, N.M., Demirci, A., Patterson, P.H., Puri, V.M., 2010. Pulsed UV light inactivation of Salmonella enteritidis on eggshells and its effects on egg quality. J. Food Prot. 73 (8), 1408–1415.

Koutchma, T., 2009. Advances in ultraviolet light technology for non-thermal processing of liquid foods. Food Bioprocess Technol. 2, 138–155.

Kowalski, W.J., 2006. Aerobiological Engineering Handbook: A Guide to Airborne Disease Control Technologies. McGraw-Hill, New York, NY.

Kuo, F., Carey, J., Ricke, S., 1997. UV irradiation of shell eggs: effect on populations of aerobes, moulds and inoculated Salmonella typhimurium. J. Food Prot. 60 (6), 639–643.

Lamikanra, O., Kueneman, D., Ukuku, D., Bett-Garber, K.L., 2005. Effect of processing under ultraviolet light on the shelf life of fresh-cut cantaloupe melon. J. Food Sci. 70, C534–C538.

Li, J., Zhang, Q., Cui, Y., Yan, J., Cao, J., Zhao, Y., et al., 2010. Use of UV-C treatment to inhibit the microbial growth and maintain the quality of yali pear. J. Food Sci. 75, M503–M507.

Lu, J.Y., Stevens, C., Khan, V.A., Kabwe, M., Wilson, C.L., 1991. The effect of ultraviolet irradiation on shelf-life and ripening of peaches and apples. J. Food Qual. 14, 299–305.

Morey, A., McKee, S.R., Dickson, J.S., Singh, M., 2010. Efficacy of ultraviolet light exposure against survival of Listeria monocytogenes on conveyor belts. Foodborne Pathog. Dis. 7 (6), 737–740.

Obande, M.A., Tucker, G.A., Shama, G., 2011. Effect of preharvest UV-C treatment of tomatoes (Solanum lycopersicon Mill.) on ripening and pathogen resistance. Postharvest Biol. Technol. 62, 188–192.

Ozer, N., Demirci, A., 2006. Inactivation of Escherichia coli O157:H7 and Listeria monocytogenes inoculated on raw salmon fillets by pulsed UV-treatment. Int. J. Food Sci. Technol. 41, 354–360.

Pombo, M.A., Rosli, H.G., Martínez, G.A., Civello, P.M., 2011. UV-C treatment affects the expression and activity of defence genes in strawberry fruit (Fragaria × ananassa, Duch.). Postharvest Biol. Technol. 59, 94–102.

Rajkovic, A., Tomasevic, I., Smigic, N., Uyttendaele, M., Radovanovic, R., Devlieghere, F., 2010. Pulsed UV light as an intervention strategy against Listeria monocytogenes and Escherichia coli O157:H7 on the surface of a meat-slicing knife. J. Food Eng. 100 (2010), 446–451.

Schalk, S., Adam, V., Arnold, E., Brieden, K., Voronov, A., Witzke, H.-D., 2006. UV-lamps for disinfection and advanced oxidation—lamp types, technologies and applications. IUVA News 8 (1).

Sommers, C.H., Geveke, D., Pulsfus, S., Lemmenes, B., 2009. Inactivation of Listeria innocua on frankfurters by ultraviolet light and flash pasteurization. J. Food Sci. 74, 3.

Sommers, C.H., Scullen, O.J., Sites, J., 2010. Inactivation of foodborne pathogens on frankfurters using ultraviolet light (254 nm) and GRAS antimicrobials. J. Food Saf. 31, 1–12.

Sosnin, E.A., Oppenländer, T., Tarasenko, V.F., 2006. Applications of capacitive and barrier discharge excilamps in photoscience. J. Photochem. Photobiol. C Photochem. Rev. 7, 145–163.

U.S. Food and Drug Administration, 2000a. 21 CFR Part 179—Irradiation in the production, processing and handling of food. Federal Register 65, 71056–71058.

U.S. Food and Drug Administration, Centre for Food Safety and Applied Nutrition, 2000b. Kinetics of microbial inactivation for alternative food processing technologies: pulsed light technology. <http://www.fda.gov/Food/ScienceResearch/ResearchAreas/SafePracticesforFoodProcesses/ucm103058.htm> (accessed 1.03.11.).

Warriner, K., Rysstad, G., Murden, A., Rumsby, P., Thomas, D., Waites, W., 2000. Inactivation of Bacillus subtilis spores on aluminium and polyethylene preformed cartons by UV-excimer laser irradiation. J. Food Prot. 63, 753–757.

Woodling, S.E., Moraru, C.I., 2005. Influence of surface topography on the effectiveness of pulsed light treatment for the reduction of Listeria innocua on stainless steel surfaces. J. Food Sci. 70, 245–351.

Yaun, B.R., Sumner, S.S., Eifert, J.D., Marcy, J.E., 2004. Inhibition of pathogens on fresh produce by ultraviolet energy. Int. J. Food Microbiol. 90, 1–8.